# Spin Pop

## Interactive Candy Toys

Unauthorized Guide for Collectors

Schiffer Publishing Ltd®

4880 Lower Valley Road, Atglen, PA 19310 USA

# Acknowledgments

I thank all those not specifically mentioned who contributed to making the Spin Pop® become reality. I would like to especially thank Debbie, for smoothing out the theft of my collection; Will and brian g smith; Deirdre (Deirdra) Gonzalez; Stacy Kaplan, for her meticulous review; Star Bucks®, for letting me visit for so long at several of their stores; and all the folks at Schiffer Publishing.


Library of Congress Catalog Card Number: 2001098862



Designed by John P. Cheek
Cover design by Bruce M. Waters
Type set in DomBold BT/Souvenir Lt BT

ISBN: 0-7643-1508-0
Printed in the United States of America

Printed in China
1 2 3 4

Published by Schiffer Publishing Ltd.
4880 Lower Valley Road
Atglen, PA 19310
Phone: (610) 593-1777; Fax: (610) 593-2002
E-mail: Schifferbk@aol.com
Please visit our web site catalog at
**www.schifferbooks.com**
We are always looking for people to write books on new and related subjects. If you have an idea for a book, please contact us at the above address.

This book may be purchased from the publisher. Include $3.95 for shipping. Please try your bookstore first.
You may write for a free catalog.

In Europe, Schiffer books are distributed by
Bushwood Books
6 Marksbury Ave. Kew Gardens
Surrey TW9 4JF England
Phone: 44 (0)20 8392-8585; Fax: 44 (0)20 8392-9876
E-mail: Bushwd@aol.com
Free postage in the UK. Europe: air mail at cost.
Please try your bookstore first.

# Contents

Dolly Lolly
Looney Tunes
Taz
Toy Story
Drill Pop
Twin Spin Pop
Double the Fun For Your Tongue!
Holiday
Easter
Stocking Stuffers

# Preface

When purchasing or collecting a Spin Pop Candy®, one gets not just candy but also a toy. With a push of a button, the spinning action comes to life by rotating a replaceable ball-shaped sucker. In addition to the rotation of a sucker, a licensed figure greets you with its personal form of salute. Since its invention in 1992, there have been many forms of this product. Most still resemble the very first handles, which had no character figures. Soon to follow was cultural branding with properties from television, movies, comics, and special promotions. Spin Pops® had the features that could make it be something big. Cap Toys, who first brought the product to market, now followed in the footsteps of a model set decades earlier in the candy business. The Spin Pop®, to many would become the "PEZ® of the nineties." The push was on to continue the success and growth of the Spin Pop® in this emerging category. This new rapidly growing sector of the confectionery industry was dubbed "interactive candy." Spin Pops® surprised many of the candy and slumbering toy companies. By combining a motorized toy and candy into one, a novel and niche market was re-discovered. In addition to being new and innovative, the Spin Pop® was introduced at a higher price than traditional candy items. This stunned the retail buyers who felt unsure of the product and thought it would not last. Years later, they continue to sell in the millions.

Collecting Spin Pop® handles is on the rise. Similar to that of the PEZ® dispensers, which were first introduced many years ago, the market for collecting Spin Pops® is growing. A benefit of this is that you can still find some of the older versions at their original retail prices. Many of these older handles may be available in the stores today, but probably will not be there very much longer. Check with local retailers in your area. You may be lucky enough to find a rare version that is no longer being made. Searching for older handles in your local stores can help start or even round out any Spin Pop® collection. Starting today, instead of later when these handles are no longer easily available, could save you money and you won't have to wait until you're older to say, "I had one of those."

Many of the early and specialty handles are hard to find because they were made in low production numbers. Be surprised if you find any handles without figures on them at retail; those most commonly found today are handles with characters and motorized features. A well-rounded collection should contain various types of handles first marketed that had no figure or license. Several of these early handles had been made for specialty stores, promotional events, and for private labeling. Other hard to find Spin Pop® handles were made exclusively for some international markets. These handles had been produced with licenses in smaller than normal quantities. International handles will be more difficult to find, but they are out there.

The resources available for collecting Spin Pops® are vast. Many of these "impulse buys" or, my favorite, "shut me up toys" still exist and are waiting to be discovered. If you search through the local store, attend a toy collector event, browse an Internet auction, or dig through that toy box at the next garage sale, you may be able to find some nice examples. You could get that missing, hard to find, or never-knew-it-existed Spin Pop® to add to your collection.

As you look through these pages, you will see examples of many production handles and packaging dating from the very first version from 1993 through to the latest at time of publication. This book is set up as a visual layout of the Spin Pop® history. You will see the very first handle to the most recent, plus examples of prototypes similar to the real thing and others that never came to life. Examples of "me too" or "knock-off" handles from competitors are also included. Other interesting works on the Spin Pop® that help tell the evolutionary story and lend credit to the development efforts are included in later sections of the book.

### About Pricing A Spin Pop®

Collecting things of interest is a passion, or even an investment, for many. Spin Pops® have a value that is roughly based on retail price plus perceived value. Specific cultural and limited production units also increase valuations. In addition, there may be a special version that means more to one individual than to another. What affects the price is the saturation and availability of any version. Demand, and the willingness to fork over the money on something that you want, changes everything. A part of this book is a price guide with the retail price, any recent auction bids, plus the actual delivery value. Since collecting Spin Pops® is a relatively young market, it is difficult to accurately define prices. There are several resources to buy and sell. In today's Internet society, finding what you want is easier than ever. This means you can locate almost anything with choices of how much to pay. Once the category matures, with your participation, information on values can better be documented.

For some, this book and the information contained may be a short trip down memory lane. For others, I hope it is an inspirational guide to the creative work and innovation that went into making the Spin Pop® come to life.

# Introduction

In 1992 a new spin on an old concept was re-invented: provide a novel candy item for the demanding consumer. The Cap Toy idea was to make a "lazy man's sucker" and sell it for a never-heard-of $2.99. Novelty and gimmick were originally the appeal of a Spin Pop®. Operating from Cleveland, Ohio, the toy manufacturer was a small entrepreneurial company focusing primarily on novelty toys. To create novelties with a gimmick at Cap Toys required searching and recruiting help from outside the company. A good way to do this was by working with inventors. Russell Gold was the person responsible for soliciting the Spin Pop® idea. He attended an invention show in Pittsburgh, Pennsylvania. There he met the inventor group B.A.A.T. Enterprises.

The folks from B.A.A.T. Enterprises at that time were two couples who had been working for the US Postal service in Virginia. Inventing for them was just a side interest. They had a vision for novel candy products, which lead to the creation of the Spin Pop®. Their journey to place the item in the market was not an easy one. Several attempts at presenting a novel candy idea had been made with many of the toy and candy companies, but to no avail. It was not until the day they met with a representative from a Cleveland company that their vision took off. Once the Spin Pop® was accepted for production at Cap Toys, a dedicated department was set up to operate the new candy side of the toy business. It was named CT Gifts and run by a small staff. The selling had begun. This group published press releases and set up booths at candy and food shows. They even hired someone to wear an inflatable suit that looked exactly like the original Spin Pop®. There even was a TV commercial made, showing the original handle and another candy item named Lazer Pop®, but it never aired.

Many of the people who first encountered the item at the shows responded positively, but the price was an issue. Lollipops generally sold for less than a dollar. In the minds of store buyers this fact would not change much, until later. Sales and buyer interest during the first year were slow, but good enough to provide revenue for continued production.

The next attempt at building the Spin Pop® into a brand was by giving it a license. Agreements were made with Turner Home Entertainment and Saban Entertainment in 1993 that led to both the Flintstones® and the Mighty Morphin Power Rangers® licenses. The first licensed handle, released early in 1994, saw Fred Flintstone's trademarked necktie and spotted orange suit pattern on a Spin Pop® label. Also released that year came the label only version of the Power Rangers® handle. Placement of the items at retail during the first couple of years was limited to a few mass retailers only.

Many of the other avenues to market, like grocery, convenience, and drug stores, had no real interest for such an expensive item. It was not until the evolution of the handle from the basic design to the dimensional figure that the large toy retailers took a chance and carried the item. With more toy appeal, retailers could place it amongst the other toys in a more toy-like setting. The Spin Pop® would now be considered more than just an expensive novelty sucker. It began to sell better than expected.

Store displays were originally designed to fit neatly on the counter. Now with a licensed molded figure attached to the handle, the counter display (CD) became larger. The very first Spin Pop® with a molded figure that increased the CD was the Batman© character, released in 1994, from DC Comics©. It increased the footprint enough that retailers considered another destination within the store. Displays moved from the counter to a shelf elsewhere in the store. This idea opened up an opportunity—place more candy items on bigger shelves and not just counters and checkout stands. Retailers began asking for more interactive candy items. To meet the demand, additional handles were in the works with more popular licensed properties. Some retailers who carried the line setup dedicated interactive candy sections within their stores.

The Spin Pop®, now next to its older cousin PEZ®, became considered the father of a toy candy revival. Many might argue that PEZ® is the grandfather of interactive candy, perhaps true in the sense of combining a licensed character to a plastic toy that has candy in it. The Spin Pop®, however, was an innovation that created new marketing and merchandising for toy and candy combinations. It brought together a category that was mainly limited to the checkout stands.

Large retailers like FAO Schwarz, started out with their New York Fifth Avenue location as a test market on the concept of a destination candy. This type of specific location within a store caught on, and the retailer made this section even bigger than anticipated. FAO even opened up additional specialty candy stores under the name FAO Schweetz. These stores look like the toy store, but sell candy instead.

The craze for candy toys was not entirely made up of the Spin Pop® or even PEZ®. It was a novel and gimmick market that grew quickly and many additional companies got into the game. The store shelves had a vast array of interactive candy items to choose from. What makes the Spin Pop® desirable and collectible is that it is branded with popular licenses. Unlike many candy items being produced and sold at retail, the Spin Pop®, like PEZ®, is a consistent year-round seller and continues to evolve with the changing times.

Cap Toys was doing so well with this new candy business that it was positioned for sale in 1994. The Russ Berry Company of New Jersey, who manufactures novelty toys and soft goods, made the purchase. Business remained as usual in Ohio with the exception of better financing and resources for future projects. The handles continued to see the release of multiple licenses and further enhancements driven by the new and aggressive parent company. Then in 1997 the company was sold once more to the Hasbro Corporation of Rhode Island. After one year of ownership, Cap Toys and Cap Candy were then relocated to Napa, California. The name Cap Candy was to remain a division under the umbrella of OddzOn. OddzOn is a toy company best known for manufacturing the Koosh® ball. Cap Toys and OddzOn together were sold to Hasbro at the same time by Russ Berry. Business decisions were then made to consolidate both companies together and move to the northern California region in 1998. Cap Candy continues to manufacture Spin Pops®.

# Chapter One
# Basic Handles

The very first Spin Pop® was designed primarily as a rounded-off, rectangular-shaped handle. Inside each handle there was room for a motor and one AA battery, plus wires and battery contact plates. On top, molded into the body, was a gearbox for torque. Positioned at a forward angle, the gearbox was tied to the motor, which rotated the pop stick holder through a small opening. This method of movement through a powered gearbox and spinning pop was the very essence of the Spin Pop®.

Every handle was surrounded by a paper label identifying product information needed for use and other consumer facts. A small die-cut hole was made in this label to access the action button. Die-cut holes are a character trait shared by every handle made. As it reads around the hole opening on the label, "press here," informs the user where to push in order to activate the spin motion. Even though the color of the action button had always been a separate color from that of the handle color, instructions were still needed where to push. Like a curious kid at heart, we see something that looks like a button and we want to touch and push it. This idea was one of the fundamental gimmicks of the Spin Pop®.

At retail, the counter display was an open box design allowing the consumer to pick up and play with the product. The very first handles were "Try Me" and came individually wrapped for protection. To make these first handles spin, you had to press the action button through the sealed bag. At first, the item was considered to be short-lived, and concern over the future of packaging was not an issue. As the product line began to grow, a blister design was adapted to protect the lollipop and to prevent pilferage. You will find more on this new packaging in later sections of this book.

Basic handles first appeared in one color scheme only, a blue body with an orange button. For the first couple of years, the basic handle did not change much in design.

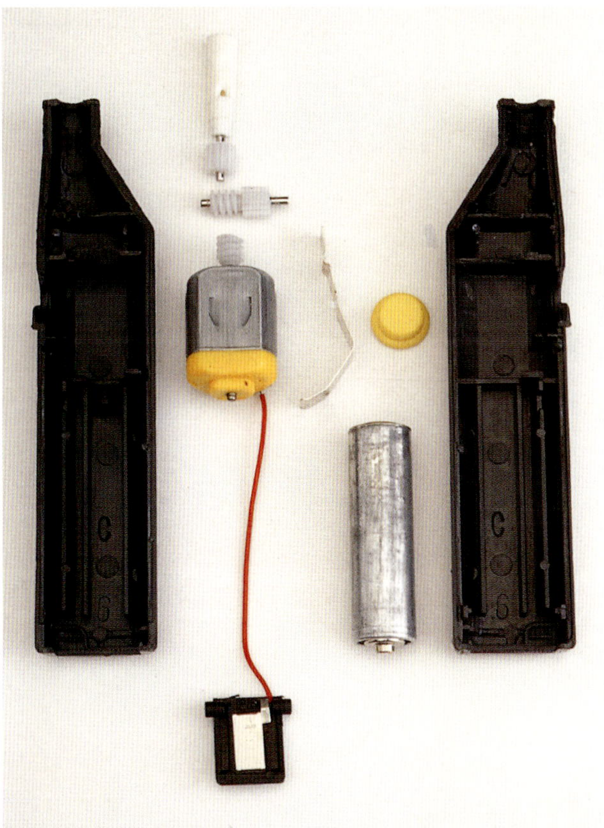

Original production handle, apart, showing internal workings.

Original handle design in blue with orange button. Shown in and out of original package with Kencraft pop. Suggested retail $2.99. First issue color scheme, Cap Toys, Ohio. 1993. $10-25

Original handle design with premium identity. Shown in all four versions made to cover birthday milestones. First ever handle to be priced higher. Suggested retail $3.99. Custom manufactured for Papel Freelance, Cranbury, New Jersey. 1993. $ 15-20

Original handle design with Power Rangers® license, Saban Entertainment, Inc©. Shown out of package with Kencraft pop. Suggested retail $2.99. Second issue color scheme, Cap Toys, Ohio. 1994. $15-25

Original handle design with Flintstones® license, Hanna Barbera Productions, Inc©. Shown out of package with Triple Treat pop. Suggested retail $2.99. First issue color scheme, Cap Toys, Ohio. 1994. $10-25

Original handle design with Biker Mice® license, Brentwood Television Funnies, Inc©. Shown out of package with Kencraft pop. Suggested retail $2.99. Second issue color scheme, Cap Toys, Ohio. 1994. $30

Original handle design with Company created Hots identity. Shown out of package with cinnamon pop and prototype label. Abnormally low production units made. Suggested retail $2.99. Second issue color scheme, Cap Toys, Ohio. 1994. $30

Original handle design with Company created Spin Breakers identity. Shown out of package with large pop. Suggested retail $2.99. Second issue color scheme, Cap Toys, Ohio. 1994. $10-15

Original handle design in purple with orange button. Shown in and out of original package with Kencraft pop. Suggested retail $2.99. Third issue color scheme, Cap Toys, Ohio. 1995. $10-25

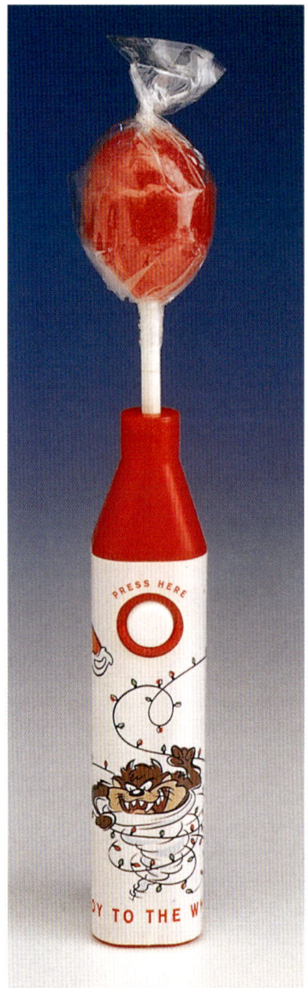


Original handle design with Warner Brothers Studio Store© labeling. This version was a premium for the holidays. Depicts Taz© entangled in a string of lights. the label reads Joy To The Whirl. Suggested retail $3.99. Custom issue color scheme, Cap

Original handle design with The Rosie O'Donnell Show© label. Packaged in a clear bag as a give away for show audience participants. Notice that the label was applied over an existing label for this premium. Various handle colors exist with this labeling practice. First issue color scheme shown, Cap Toys, Ohio. 1995. $10-20.

Original handle design with Bennigan's restaurant premium label for the Blarney Blast. This is a Saint Patrick's Day party give away. Shown in and out of original package with Kencraft pop. First issue color scheme, Note on this label the dates indicate 1993 and 1995. The release of this item was for the later. Cap Toys, Ohio. 1995. $10-25

New and improved Spin Pop® handles were on the drawing table. The basic handle evolved into a dimensional character, and it was decided that all future handles would be made this way. The innovations applied to the Spin Pop® after the decision to.use molded characters came in the form of cloth or articulated features. Figures would then have movement through the handle's internal motorized system. After the Power Ranger® and Flintstones® label-only versions, Batman© was the first handle with a dimensional figure. It boasted a rigid half torso pose, but attached was a dynamic cloth cape.

The use of cloth was an enhancement also made on one of Cap Toys' own creations for girl consumers. The Dolly Lolly®, released in 1995, was a handle that not only had a cloth skirt, but also was the first company creation to have the motorized spin motion. This spinning motion was then applied to the Warner Brothers Studios© Looney Tunes® Taz© character released later in 1995. This became the very first licensed handle to feature a motorized spinning figure and manually posed appendages. All handles with dimensional figures originated from the Batman© version. Articulation of appendages and motorized movement was inspired by, and can trace its roots to, the Taz© character.

Dolly Lolly®, a company creation. Rotating figure. Shown in and out of package with Spangler Triple Treat pop. First girl specific handle made. Notice special label location to the side of figure. This was in order so the button would not appear in center of skirt. Suggested retail $3.99. Cap Toys, Ohio. 1995. $3-5

Taz©, Looney Tunes®, and Warner Bros©. Rotating figure. Shown in and out of package with Charms Blow pop. First blister pack to feature a hole for retail peg. Suggested retail $3.99. Cap Toys, Ohio. 1995. $3-10

Future articulation and motorized motion would be seen on nearly every new handle to follow. If a Spin Pop® was to have an articulated figure, complete or partial, motion was either motorized or manual. Many other licenses found their way onto handles, which presented new challenges in movement. Each new figure would present a challenge specific to that characters personality. In some situations, the licenser did not allow partial figures.

Special markets, such as Canada and the Disney© theme parks, had the first handles to feature full-bodied figures that had motion. The candy store at the Disney World© theme park in Florida and some Canadian markets had, at the time of introduction, exclusive handles in limited production numbers. For Florida, there was a Pooh® character with up-and-down arm motion, and a Tigger® figure. The Tigger® was a complete figure including a coiled tail that featured an upward bounce motion and a first ever rearward-facing action button. Canada saw a trio of the most popular Disney© classic characters. Canadian handles had spinning bodies or a waving arm motion. Some of these special market handles have since been supplied to the US mass retail with new color schemes and labels.

Tigger©, Disney©. Spinning body motion. Shown in and out of package with Chupa Chups pop. Premium market product exclusive for Canada. Note, label does not have year printed. OddzOn, Ohio. 1998. $20-25

Mickey Mouse©, Disney©. Up and down arm motion. Shown in and out of package with Chupa Chups pop. Premium market product exclusive for Canada. Note, label does not have year printed. OddzOn, Ohio. 1998. $20-25

Pooh©, Disney©. Spinning umbrella motion. Shown in and out of package with Chupa Chups pop. Premium market product exclusive for Canada. Note, label does not have year printed. OddzOn, Ohio. 1998. $5-15

More changes to the handle have occurred and continue to do so since Hasbro took over; more so than in the Cap Toys or Russ Berry years combined. Many of these enhancements to the handles have been for performance and/or cost-cutting reasons. The most physical change to the handle, excluding the switch to a dimensional character, happened during this ownership. The first notable advances, the ones as a collector you should have a basic knowledge of, happened during the Cap Toys years between 1992 and 1997. They represent some of the biggest leaps forward in mechanisms and creativity towards expanding the line. Most of these early mechanical advances are still being used in today' s handles. What changes you can see currently come in the form of new characters from popular licenses. All the changes, by Cap Toys, Russ Berry, or Hasbro, represent a variety of improvements to the handles.

One thing that happened was the addition of another factory in China. Two factories were now producing Spin Pops® for Cap Toys between 1992 and 1997. After 1997, the original factory assigned to produce handles was dropped. Every factory in China has its own way of doing business and mold-making procedures. Because of this, you can identify small variances in the handle shape and size from one factory to another. There is very little visible difference between the two factories' products.

As a collector, you should be aware of this, but don' t try and determine a value to the subtleties, there is none. Differences you can watch for are among the handles with characters introduced later. You will be able to identify these handles in the following chapters. Basic handles had only new plastic colors and different labels during the first two years.

Other attempts at expanding the basic line included themed candy, premiums, and licensed approaches. These changes brought about flavors such as red-hot cinnamon, two miniature grape, cherry pops, plus a mouth-stuffing jawbreaker. Other than color, new packaging, and display boxes, Spin Pops® would remain as they were until the next generation of handles with dimension.

# Chapter Two
# Dimensional Figures

Licensing of handles by use of labels only, outside of special markets, ended primarily in 1994 with the introduction of Batman©. Spin Pops® then took that turn towards the toy world. By adding the dimensional figure with licensed characters, handles became less of a gimmick and more a toy collectible. 1994 was the year that new handles had the dimensional character. The White Ranger©, although released in 1995, and Batman© became the very first of many handles to follow that featured the first ever half-torso character. In the years to come, the world would see many new releases for the Spin Pop® with articulation and motorized motion.

Batman©, DC Comics©. First ever handle to feature a half torso figure and cloth cape. Two versions exist. One with upward and one with downward mouth. Upward mouth Shown in and out of package with Spangler Triple Treat pop. Notice the blister pack has no hole for retail peg. Suggested retail $3.99. Cap Toys, Ohio. 1994. $15-20

White Ranger©, Power Rangers®, Saban Entertainment, Inc©. Shown in and out of package with Spangler Triple Treat pop. Notice the blister pack has no hole for retail peg. Suggested retail $3.99. Cap Toys, Ohio. 1995. $20-30

Between the years 1994 and 1996, Cap Toys was making a transition to dedicating more resources to the candy effort. During this period, the Spin Pop® was only a part of the Cap Toys line launched in the interactive candy category. Development of Spin Pops® during this time frame consisted of existing product and other variations on the spinning mechanism. Many of the new candy products that followed also used a rotating lollipop like the Spin Pop®. These various themes were created to diversify the Cap Candy line. You will see examples of spinning variations in later sections of this book. Product development at this time was focused and driven to build relationships with licensers. Agreements with Time Warner® allowed the first licensed figures that had motorized movement. Taz©, released late in 1995, had moveable limbs and a motorized spinning body. Also developed during this period, but released in early 1996, was Bugs Bunny© and Marvin the Martian©. They were the first handles to feature motorized movement of a body part. Both figures touted a moving right arm holding one of the character icons. Also a first time license with Disney© provided for the timely release of the popular Buzz Lightyear© character. It featured manually operated wings only. This period would also produce Spin Pops® with static figures, as well as non-motorized, and motorized articulated handles. Other non-licensed company creations had been under development as well.

Marvin the Martian©, Looney Tunes®, Warner Bros©. Up and down arm motion. Shown in and out of package with Spangler Triple Treat pop. Suggested retail $3.99. Cap Toys, Ohio. 1996. $3-10

Bugs Bunny©, Looney Tunes®,and Warner Bros©. Up and down arm motion. Shown in and out of package with Spangler Triple Treat pop. Suggested retail $3.99. Cap Toys, Ohio. 1996. $3-10

Buzz Lightyear©, Toy Story®, Disney©. Manual wing motion. Shown in and out of package with Chupa Chups pop. First handle to feature a branded candy. Note, this label does not have a year printed on it. First issue labeling. Suggested retail $3.99. Cap Toys, Ohio. 1996. $5-10

They were designed for two of the four holiday seasons. At retail, two Xmas "stocking stuffers" and a pair of Easter bunnies made it on the shelf. For Xmas there was a Santa Claus with a bag of goodies and a top hat-wearing Snow Man figure. Both came without movement of any kind. Easter brought about two colorful hat and ear twirling rabbits holding a basket of eggs. For the girls market specifically was the Dolly Lolly®. This was the first non-licensed handle to feature a spinning character.

Snow Man, Company creation. Shown in and out of package with Spangler Triple Treat pop. Seasonal released item, low production. Suggested retail $3.99. Cap Toys, Ohio. 1996. $5-15

White Easter bunny with spinning hat and ears. Company creation. Shown in and out of package with Spangler Triple Treat pop. Seasonal released item, low production. Suggested retail $3.99. Cap Toys, Ohio. 1996. $5-15

Yellow Easter bunny with spinning hat and ears. Company creation. Shown in and out of package with Spangler Triple Treat pop. Seasonal released item, low production. Suggested retail $3.99. Cap Toys, Ohio. 1996. $5-15

Santa Claus, Company creation. Shown in and out of package with Spangler Triple Treat pop. Seasonal released item, low production. Suggested retail $3.99. Cap Toys, Ohio. 1996. $5-15

Many new products have been released from the licensing efforts. Company creations were introduced almost every year. These were sold at a lower retail price than the licensed units. In addition, they targeted a specific audience that the licenses did not. New handles with figures would become either licensed or seasonal, with an exception in the girls market. Girls line extensions made by the company include the Unicorn Pop, released in 1997, and the Princess Pop® in 1999. The Unicorn Pop® brought about the first handle with hair play. This unit featured a multi-colored vertical spiral pop. The same candy piece inspired another company creation, this time targeted for the boys market.

Unicorn Pop®, Company creation. Up and down leg motion. Shown in and out of package with Candycolor Unicorn pop. Three color schemes of white, blue and purple shown. Suggested retail $3.99. Cap Toys, Ohio. 1996. $5-10

Robin©, The Adventures of Batman and Robin®, DC Comics©. Shown in and out of package with Chupa Chups pop. Suggested retail $3.99. Cap Toys, Ohio. 1997. $5-15

Batman©, The Adventures of Batman and Robin®, DC Comics©. Shown in and out of package with Chupa Chups pop. Same figure as first released with new label. Suggested retail $3.99. Cap Toys, Ohio. 1997. $5-10

Tigger©, Winnie the Pooh®, Walt Disney Attractions, Inc©. Shown in and out of package with Chupa Chups pop. Up and down bounce motion. Premium market product exclusive for the Florida theme parks. Sold in bulk for use on retail peg. This was the first time of a rearward facing action button. This feature was the first big step towards future changes and was fought then with opposition. Note, label does not have year printed. Suggested retail $5.99. Cap Toys, Ohio. 1997. $10-15

Ariel©, The Little Mermaid®, and Disney©. Back and forth tail section motion. Shown in and out of package with Chupa Chups pop. Largest pattern made at time. Developed for release with home video. Longer than expected licenser process called for running change of pattern. New character pattern to be in effect after first 100K shipped. Note, label does not have year printed. Suggested retail $3.99. OddzOn / Cap Toys, Ohio. 1997. $20-25

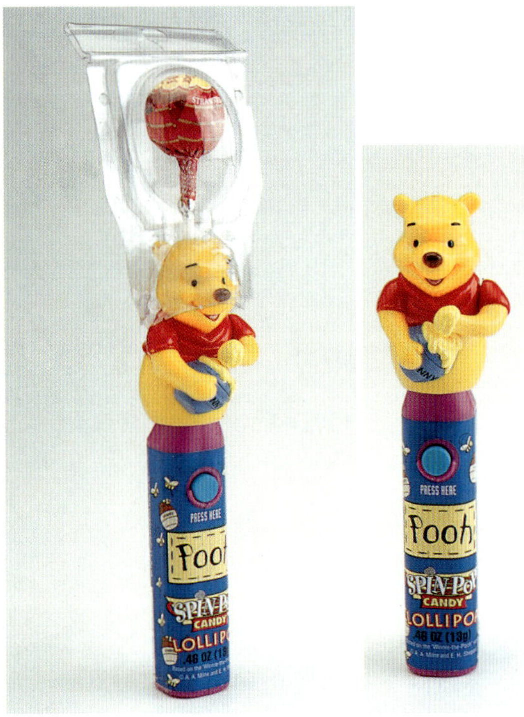

Pooh©, Winnie the Pooh®, Walt Disney Attractions, Inc©. Shown in and out of package with Chupa Chups pop. Up and down arm motion. Premium market product exclusive for the Florida theme parks. Sold in bulk for use on retail peg. Note, label does not have year printed. Suggested retail $5.99. Cap Toys, Ohio. 1997. $15-20

Ariel©, The Little Mermaid®, and Disney©. Back and forth tail section motion. Shown in and out of package with Chupa Chups pop. Revised pattern made as running change. Note, label does not have year printed. Suggested retail $3.99. OddzOn /Cap Toys, Ohio. 1997. $10-15

Mickey Mouse©, Disney©, Walt Disney Attractions, Inc©. Shown in and out of package with Chupa Chups pop. Up and down arm motion. Premium market product exclusive for the Florida theme parks. Sold in bulk for use on retail peg. Note, label does not have year printed. Suggested retail $5.99. OddzOn /Cap Toys, Ohio. 1997. $10-15

Sylvester© and Tweety©, Looney Tunes®, Warner Bros©. Rotating disc and up and down arm motion. First time for dual characters with motion combined on a handle. Shown in and out of package with Chupa Chups pop. Suggested retail $3.99. OddzOn /Cap Toys, Ohio. 1998. $3-10

Minnie Mouse©, Disney©, Walt Disney Attractions, Inc©. Shown in and out of package with Chupa Chups pop. Up and down arm motion. Premium market product exclusive for the Florida theme parks. Sold in bulk for use on retail peg. Note, label does not have year printed. Suggested retail $5.99. OddzOn /Cap Toys, Ohio. 1997. $10-15

Chip©, Small Soldiers®, Universal City Studios, Inc©. Motorized up and down with manual arm motion. Shown in and out of package with Chupa Chups pop. Suggested retail $3.99. OddzOn, Ohio. 1998. $5-15

Archer©, Small Soldiers®, Universal City Studios, Inc©. Motorized up and down with manual arm motion. Shown in and out of package with Chupa Chups pop. Suggested retail $3.99. OddzOn, Ohio. 1998. $5-15

Godzilla©, Toho Co., Ltd. TriStar Pictures, Inc©. Open and close mouth plus manual arm motion. Shown in and out of package with Chupa Chups pop. Suggested retail $3.99. OddzOn, Ohio. 1998. $3-10

The years from 1997 to the present brought about new handle characters and candy. Milestones in production can be seen with the introduction of name brand candy suckers. The lollipop being perishable and temporary; it is not considered a valuable part of the toy's collectible interest. For discussion purposes, you should be aware of the manufacturers who supplied "fill" for the Spin Pop®. The first branded switch in candy happened with the introduction of a Chupa Chups pop in 1996. This was done with the idea that a name brand sucker would help sell the product better. These new pops first appeared with the Buzz Lightyear© handle. All candy suppliers are identified in their order of appearance:

Kencraft, 1993, made the grape, cherry, and sweet and sour pop
Spangler, 1994, made the Triple Treat with lime, lemon, and orange, and a custom cinnamon pop
Chupa Chups, 1996, supplied pops in strawberry and grape
Candycolor, 1996, made the spiral shaped Unicorn Pop in multiple flavors
Nabisco, 1998, provided the roll of Life Savers.
Tootsie Roll, 1999, who makes the Charms Brand, supplied the Blow Pop
Foreign Candy, 1999, provided the Mega Warheads Soft Sour Bubble Gum filled pop
Joyco Confections, 2000, supplies a custom label pop in strawberry flavor

Princess Dot©, A Bug's Life®, Disney /Pixar©. No movement of character. Shown in and out of package with Chupa Chups pop. First mass retail handles to feature a fully dimensional and licensed figure molded separately. Note, label does not have year printed. Suggested retail $3.99. OddzOn, California. 1998. $10-15

Flik©, A Bug's Life®, Disney /Pixar©. No movement of character. Shown in and out of package with Chupa Chups pop. First mass retail handles to feature a fully dimensional and licensed figure molded separately. Note, label does not have year printed. Suggested retail $3.99. OddzOn, California. 1998. $10-15

Chuckie©, Rugrats®, Viacom International Inc©. Motorized arm and jaw motion. Shown in and out of package with Chupa Chups pop. First time a label was printed bilingual for both domestic and foreign markets. Suggested retail $3.99. First issue labeling. OddzOn, California. 1998. $10-15

Tommy©, Rugrats®, Viacom International Inc©. Motorized and manual arm movement. Shown in and out of package with Chupa Chups pop. First time a label was printed bilingual for both domestic and foreign markets. Suggested retail $3.99. First issue labeling. OddzOn, California. 1998. $10-15

Princess Pop®, a company creation. Motorized up and down arm motion. Shown in and out of package with Chupa Chups pop. OddzOn, California. 1998 & 1999. $5-10

Tarzan & Sabor©, Tarzan®, Disney©. No motion. First time two fully dimensional characters appeared together. Shown in and out of package with first time appearance of the Charms pop. Suggested retail $3.99. OddzOn, California. 1999. $10-15

Qui-Gon Jinn©, Star Wars Episode One®, Lucasfilm Ltd©. Up and down saber chop motion. Shown in blister card package with Joyco pop. First ever handle and figure with this type of package. Suggested retail $5.99. OddzOn, California. 1999. $5-10

Terk©, Tarzan®, Disney©. No motion. Shown in and out of package with first time appearance of the Charms pop. Suggested retail $3.99. OddzOn, California. 1999. $10-15

Woody©, Toy Story 2®, Disney /Pixar©. No motion. Shown in and out of package with Charms pop. Suggested retail $3.99. OddzOn, California. 1999. $5-10

Darth Maul©, Star Wars Episode One®, Lucasfilm Ltd©. Up and down saber chop motion. Shown in blister card package with Joyco pop. First ever handle and figure with this type of package. Suggested retail $5.99. OddzOn, California. 1999. $5-10

Buzz Lightyear©, Toy Story 2®, Disney /Pixar©. Manual wing motion. Shown in and out of package with Charms pop. Second issue labeling. Suggested retail $3.99. OddzOn, California. 1999. $5-10

Wally Warhead©, The Original Mega Warheads®, The Foreign Candy Company, Inc.©. In and out facial motion. Shown in and out of package with Soft Sour pop. Suggested retail $3.99. OddzOn, California. 1999. $5-10

Pooh©, Winnie the Pooh®, Disney©. Spinning umbrella motion. Shown in and out of package with Charms pop. Former premium market item released for mass retail. Suggested retail $3.99. OddzOn, California. 1999. $10-15

Snoopy© and Woodstock©, Peanuts®, United Feature Syndicate, Inc©. Spinning disc motion. Shown in and out of package with Charms pop. Suggested retail $3.99. OddzOn, California. 1999. $5-10

Tigger©, Winnie the Pooh®, Disney©. Up and down bounce motion. Shown in and out of package with Charms pop. Former premium market item released for mass retail. Suggested retail $3.99. OddzOn, California. 1999. $5

Carnitore©, Dinosaur®, and Disney©. Open and close jaw motion. Shown in and out of package with Charms pop. Suggested retail $3.99. OddzOn, California. 2000. $5-15

Aladar©, Dinosaurs®, and Disney©. Open and close jaw motion. Shown in and out of package with Charms pop. Suggested retail $3.99. OddzOn, California. 2000. $5-15

Pikachu©, Pokemon®, and Nintendo©. No mechanized motion. Shown in and out of package with Charms pop. Suggested retail $3.99. OddzOn, California. 2000. $5-10

Pooh and Tigger©, Winnie the Pooh®, Disney©. Original premium market duo that was the first ever-classic character assortment made. Cap Toys, Ohio. 1997.

Squirtle©, Pokemon®, and Nintendo©. No mechanized motion. Shown in and out of package with Charms pop. Suggested retail $3.99. OddzOn, California. 2000. $5-10

Undertaker©, World Wrestling Federation®, WWF Entertainment, Inc©. Up and down arm and chair motion. Shown in and out of package with Charms pop. Released in 2000. Suggested retail $3.99. OddzOn, California. 1999. $5-10

Shaggy and Scooby®, Scooby Doo®, Cartoon Network©. Rapid vibrating leg and character motion. Shown in and out of package with Charms pop. Suggested retail $3.99. OddzOn, California. 2000. $5-8

Stone Cold©, World Wrestling Federation®, WWF Entertainment, Inc©. Up and down arm and motion. Shown in and out of package with Charms pop. Released in 2000. Suggested retail $3.99. OddzOn, California. 1999. $5-10

Power Puff Girls© grouping, Cartoon Network©, Warner Bros©. OddzOn, California. 2000.

Blossom® Power Puff Girls©, Warner Bros©. Up and down bounce motion. Shown in and out of package with Charms pop. Suggested retail $3.99. OddzOn, California. 2000. $5

Bubbles® Power Puff Girls©, Cartoon Network©. Up and down bounce motion. Shown in and out of package with Charms pop. Suggested retail $3.99. OddzOn, California. 2000. $5

Buttercup® Power Puff Girls©, Cartoon Network©. Up and down bounce motion. Shown in and out of package with Charms pop. Suggested retail $3.99. OddzOn, California. 2000. $5

Xmas Holiday stocking stuffer. Up and down bounce motion of figurehead. Shown in and out of package with Charms pop. First appearance of different shaped action button and a fully characterized handle body. Suggested retail $3.99. OddzOn, California. 2000. $5-7

Angelica©, Rugrats / Nick Toons®, Viacom International, Inc©. Up and down arm motion with posed ponytails. Shown in and out of package with Charms pop. Released in 2001. Suggested retail $3.99. OddzOn, California. 1999. $5

Nick Toons© grouping, Cartoon Network©, Warner Bros©. OddzOn, California. 2001.

Baby Dill©, Rugrats / Nick Toons®, Viacom International, Inc©. Motorized in and out face and pacifier motion. Shown in and out of package with Charms pop. Released in 2001. Suggested retail $3.99. OddzOn, California. 1999. $5

Tommy©, Rugrats / Nick Toons®, Viacom International, Inc©. Motorized and manual arm movement. Shown in and out of package with Charms pop. Second issue labeling. re-released in 2001. Suggested retail $3.99. OddzOn, California. 1999. $5

Chuckie©, Rugrats / Nick Toons®, Viacom International, Inc©. Motorized arm and jaw motion. Shown in and out of package with Charms pop. Second issue labeling. re-released in 2001. Suggested retail $3.99. OddzOn, California. 1999. $5

Sponge Bob Square Pants©, Nickelodeon / Nick Toons®, Viacom International, Inc©. Motorized up and down pant and arms motion. Shown in and out of package with Charms pop. Suggested retail $3.99. OddzOn, California. 2000. $5

Wolverine®, X Men®, Marvel Characters, Inc©. Motorized up and down arm motion. Shown in and out of package with Charms pop. Suggested retail $3.99. OddzOn, California. 2000. $8

Cat Dog©, Nickelodeon / Nick Toons®, Viacom International, Inc©. Motorized dual pop spin motion. Shown in and out of package with two Charms mini pops. First ever Twin Spin Pop® with license. Suggested retail $3.99. OddzOn, California. 2000. $5

Magneto®, X Men®, Marvel Characters, Inc©. Motorized up and down arm motion. Shown in and out of package with Charms pop. Suggested retail $3.99. OddzOn, California. 2000. $8

Action Man©, Dr. X®, Hasbro International Inc©. Motorized in and out arm motion. Shown in and out of package with Charms pop. First appearance of clear blister with Cap Candy molded in. Domestic and International labels exist. Suggested retail $3.99. OddzOn, California. 2000. $5

Action Man©, AM®, Hasbro International Inc©. Motorized head and arm motion. Shown in and out of package with Charms pop. First appearance of a clear blister with Cap Candy molded in. Domestic and International labels exist. Suggested retail $3.99. OddzOn, California. 2000. $5

Easter, bunny trio in assorted colors. Motorized up and down motion of eggs in basket. Shown in and out of package with Charms pop. First appearance of the information tag attached. Suggested retail $3.99. OddzOn, California. 2000. $5

Manchester United©, Hasbro International, Inc©. Group shot. Hasbro, UK. 2000.

Beckham #7, Manchester United©, Hasbro International, Inc©. Motorized head and arm motion. Shown in and out of package with Joyco pop. Suggested retail $3.99. Hasbro, UK. 2000. $15

Yorke # 19, Manchester United©, Hasbro International, Inc©. Motorized head and arm motion. Shown in and out of package with Joyco pop. Suggested retail $3.99. Hasbro, UK. 2000. $15

Giggs # 11, Manchester United©, Hasbro International, Inc©. Motorized head and arm motion. Shown in and out of package with Joyco pop. Suggested retail $3.99. Hasbro, UK. 2000. $15

Tweenies©, Tell-Tale Productions Ltd., BBC Television©. Rotating figure motion. Shown in and out of package with Joyco pop. Suggested retail $3.99. Hasbro, UK. 2000. $15

Rocky©, Chicken Run®™&© DreamWorks, L.L.C., Aardman Chicken Run and Pathé Image. In and out knee motion with flexible face feature. Shown in and out of package with Joyco pop. Suggested retail $3.99. Hasbro, UK. 2000. $15

Ginger©, Chicken Run®™&© DreamWorks, L.L.C., Aardman Chicken Run and Pathé Image. In and out knee motion with flexible face feature. Shown in and out of package with Joyco pop. Suggested retail $3.99. Hasbro, UK. 2000. $15

Minnie Mouse©, Disney©, Rotating figure motion. Shown in and out of package with Joyco pop. First appearance of the new square handle design. Suggested retail $3.99. OddzOn, California. 2000. $5

Mickey Mouse©, Disney©, Rotating figure motion. Shown in and out of package with Joyco pop. First appearance of the new square handle design. Suggested retail $3.99. OddzOn, California. 2000. $5

T-Rex, Jurassic Park 3®, Universal Studios© and Amblin Entertainment, Inc©. Motorized open and close jaw motion. Shown in and out of package with Joyco pop. Suggested retail $3.99. OddzOn, California. 2001. $5

Spinosaurus, Jurassic Park 3®, Universal Studios© and Amblin Entertainment, Inc©. Motorized open and close jaw motion. Shown in and out of package with Joyco pop. Suggested retail $3.99. OddzOn, California. 2001. $5

Pooh©, Winnie the Pooh®, Disney©. Motorized up and down arm motion. Shown in and out of package with Joyco pop. Third revised character to be assorted with bouncing Tigger©. Suggested retail $3.99. OddzOn, California. 2001. $5

# Chapter Three
# Store Displays

Spin Pops® first releases came in a cardboard display unit referred to as a counter display (or CD) that held 24 individually wrapped handles. This display box also functioned as part of the shipping container by securing the handles for transit. The CD had to be removed from the outer shipping container. When the CD was to be set up, the packing material was removed and the box top was folded back. This folding back would separate the die-cut header card from the box top and the card was to be inserted behind the handles for display.

Original handles were shipped in master packs of two 24-count display boxes. After the change to the dimensional characters, the counter display also changed. Box dimensions kept increasing due to the addition of new and different dimensional characters. Attempts to standardize the overall size were made. Generally though, if the product was bigger or wider, it was because that was what the mechanism required for that licensed character to move. Some complaints had been made from retailers about the increasing CD size. In order to maintain the 12 pieces per CD desired by retailers, the boxes had to, and did, grow in size. Mass retail would soon grow accustomed to the fact that the display box got bigger. With increased sales numbers and specific locations now within many stores, the complaints subsided.

To further address the issue of new in-store locations, and the increasing display size, Cap Candy was looking for new ways to merchandise the Spin Pop®. One of the first concerns was how to position the handle elsewhere within the store. Clear outer wrappings, such as bags and plastic blisters, had been used on early handles to protect the sucker. One easy solution was to provide a small hole for hanging. Both the clear bag and outer blister would be altered to accommodate a round hole for the retail peg. Every handle shipped, after the introduction of Taz©, would be designated for use either in a minimum 12-count CD or as an individually pegged item anywhere within the store. Handles could now be sold to certain retailers, if desired, as a bulk item without a costly cardboard display box.

Early merchandising plans called for purpose-built units. These cardboard displays were sold to retailers on specific properties. For example, the Looney Tunes® line was developed into three types of temporary store fixtures. The sidekick, with either a two or four level plan, would hang neatly on the edge of a store rack. Also, a stand-alone floor model was shipped. It housed product vertically in a reclined unit that had some assembly required to be upright. All intended as temporary units and shipped filled with product. The handle quantity shipped in a special display was higher than a single CD. All three formats had their own special graphics on the display. Some would be branded with artwork for a specific license or property. With the growing candy category, Cap Candy would next develop an in-store metal rack system. These units could accommodate both the single 12-count CD and several individually pegged items. These rack systems are commonly found at Toys R Us stores.

The following images represent most of the displays produced from 1992 to the present. Only a few of the very early display boxes are shown, but enough examples to give you a taste of the general box design and construction. Overall, box shape and design remained relatively the same for many years.

While no prices are included under the captions in this section, as a guide you might expect to pay for an entire display box full of product at the going rate for that handle multiplied by the quantity. If the CD is a stand alone item for sale, expect to purchase it in the range of five dollars for a rough example up to fifteen dollars for a really nice one. Generally, most retail stores will part with the display boxes when they are empty or if you ask nicely.

Original Spin Pop®. Company creation with no dimensional figure. Display with handle in third color scheme of purple. Shown in 24-count CD with Spangler pop.

Spin Breakers®. Company creation with no dimensional figure. Display without handles. Shown in 24-count CD.

Power Rangers®. Display without handles. Note this is a Lazer Pop© box. Graphic layout is exact to the Spin Pop© CD, except for wording.

Twin Spin Pop®. Company creation with no dimensional figure. Display without handles. Shown in 12-count CD.

Unicorn Pop®. Company creation shown in 12-count CD with spiral shaped Unicorn pop.

Dolly Lolly®. First company creation with dimensional figure. Shown in 12-count CD with Chupa Chups pop.

Holiday. Company creation for Xmas season with two figures together. Shown in 12-count CD with Chupa Chups pop.

Easter. Company creation for Easter season with two figures together. Shown in 12-count CD with Chupa Chups pop.

Princess Pop®. Company creation for the girls market. Shown in 12-count CD with Chupa Chups pop.

Drill Pop®. Company creation for the boy market in two color schemes. Shown in 12-count CD with second issue labeling and spiral shaped Unicorn pop.

Easter. Company creation for Easter season with three fully dimensional figures together. Shown in 12-count CD with Charms pop.

Holiday. First company creation for Xmas season to feature the fully dimensional figure handle. Shown in 12-count CD with Charms pop.

The Adventures of Batman and Robin®. Licensed CD with both the old Batman© and new Robin© figures together. Shown in 12-count CD with Chupa Chups pop.

Toy Story®. Licensed display with Buzz Lightyear© figure. Shown in 12-count CD with Chupa Chups pop.

Small Soldiers®. Licensed display with both Chip© and Archer© figures together. Shown in 12-count CD with Chupa Chups pop.

The Little Mermaid®. Licensed display with Ariel© figure. Shown in 12-count CD with Chupa Chups pop.

Mickey for Kids®. Internationally licensed display with top hat Mickey©. Shown in 12-count CD with Chupa Chups pop.

A Bug's Life®. Licensed display with both Flik© and Dot© figures together. Shown in 12-count CD with Chupa Chups pop. The assortment had only 4 Flik© figures shipped.

Tarzan®. Licensed display with both Tarzan© and Turk© figures together. Shown in 12-count CD with Charms pop.

Dinosaur®. Licensed display with both Carnitore© and Aladar© figures together. Shown in 12-count CD with Charms pop. First appearance of the new flat CD design.

Toy Story 2®. Licensed display with both the former Buzz Lightyear© and Woody© figures together. Shown in 12-count CD with Charms pop.

Winnie the Pooh®. Licensed display with both Tigger© and Pooh© figures together. Shown in 12-count CD with Charms pop. New color schemes for both handles. First time a CD combined handles from former international and premium markets.

Winnie the Pooh®. Licensed display with both Tigger© and Pooh© figures together. Shown in 12-count CD with Joyco pop. New dimensional figure for the Pooh© handle.

Mickey and Minnie Mouse®. Licensed display with both Mickey© and Minnie© figures together. Shown in 12-count CD with Joyco pop. New version of the handle featuring a smaller figure and backdrop.

Taz©. Licensed display with Taz© figure. Shown in 12-count CD with Chupa Chups pop. Original release display had Spangler pop.

Looney Tunes®. Licensed display with both the Bugs Bunny© and Marvin the Martian© figures together. Shown in 12-count CD with Chupa Chups pop.

Looney Tunes®. Licensed display with Bugs Bunny©, Marvin the Martian©, Taz©, Sylvester© and Tweety© figures all together for the first time in one CD. Shown in 12-count CD with Joyco pop.

Sylvester & Tweety®. Licensed display with both the Sylvester© and Tweety© figures together on same handle. Shown in 12-count CD with Chupa Chups pop.

The Powerpuff Girls®. Licensed display with all three Buttercup©, Bubbles©, and Blossom© figures together. Shown in 12-count CD with Joyco pop.

Rugrats®. Licensed display with both Tommy© and Chuckie© figures together. Shown in 12-count CD with Chupa Chups pop.

Scooby Doo®. Licensed display with all Shaggy© and Scooby© figures together on the same handle. Shown in 12-count CD with Joyco pop.

Nickelodeon®. Licensed display with all new Nicktoons© characters Angelic©, Baby Dill©, Sponge Bob©, and Cat Dog© figures together with the two Tommy© and Chuckie© from previous. Shown in 12-count CD with Joyco pop.

Mega Warheads®. Licensed display with Wally Warhead© character. Shown in 12-count CD with Mega Warheads® pop.

Godzilla®. Licensed display with main character from movie. Shown in 12-count CD with Chupa Chups pop.

Pokemon®. Licensed display with both Pikachu© and Squirtle© characters. Shown in 12-count CD with Charms pop.

Snoopy®. Licensed display with both Snoopy© and Woodstock© characters together one the same handle. Shown in 12-count CD with Charms pop.

WWF®. Bulk shipper with no display. Each box contained six figures for the retail hook. Only two mass retailers received this method of shipping. All others received full CD cases. Shown with Charms pop.

WWF®. Licensed display with both Undertaker© and Stone Cold© characters together. Shown in 12-count CD with Charms pop.

Chicken Run®. Licensed display with both Rocky© and Ginger© characters together. Shown in 12-count CD with Joyco Pop.

Action Man®. Licensed display with both AM© and Dr. X© characters together. Shown in 12-count CD with Charms pop. Shipped in both bulk and CD formats.

Jurassic Park 3®. Licensed display with both a Spinosaurus© and T-Rex© figures together. Shown in 12-count CD with Joyco pop.

Freestanding floor unit. Pre-packed 48-count licensed display with both Bugs Bunny© and Marvin the Martian© figures together.

Sidekick unit. Pre-packed 32-count licensed display with all three Bugs Bunny©, Marvin the Martian©, and Taz© figures together.

End-cap unit. Pre-packed 64-count licensed display with all three Bugs Bunny©,Marvin the Martian©, and Taz© figures together.

Destination Candy. Planogram example of an entire end-cap full of Spin Pops®. Notice the display box with original blue handle in center.

# Chapter Four
# On the Inside

The complete production schedule for making a Spin Pop® was generally about five months. Factor in a lengthy licenser approval and it could increase by another month or two. Selling licensed handles required early concept and creative work to gain approvals. Often the sales force would make calls on buyers to preview a new licensed property. In the early years, the Spin Pop® was so new that these visits were best served with a tangible selling aid. Sales efforts would demand a mock-up or prototype sample in order to better sell to retail buyers. It also served as an early way of gauging if a property had potential to be successful. Internal line reviews would also provide for some of the most rare and unique Spin Pops® ever fabricated. Line review meetings were set up to review the next wave of products to be released. Often handmade or quickly sculpted models would be generated for this purpose. Many of the concept handles made would be reproduced in multiples until actual tooling was started. Samples, made of rapid prototype materials, would be decorated and labeled to look like the real thing. Many samples would also have functional features, others not at all. Brokers and sales people often would receive a mock-up sample of new handles for their needs. It is likely that many samples still exist, waiting to be re-introduced into the market as a rare collectible.

In this chapter, images of ultra-rare and one-of-a-kind Spin Pop® prototypes are presented, the work of inventors, Cap employees, and competitors. The information and images are samplings from the category and outlined as a foundation for the collector to reference. Some of the defining work of the Spin Pop® is presented and identified. Technical details are shown as part of the evolutionary steps taken. They will help collectors identify variances and provide other information on handles for any collector level.

## Prototypes

The inventors of the Spin Pop® concept had to come up with a tangible way of showing their idea to the world. Working with found and store bought materials, they fashioned together a working model. By combining plastic and metal together with glue and paint, the inventors made an idea come to life. This is reminiscent of the same process, which generated many of the mock-ups at Cap Candy. Cap's modeling resources were both inside and outside the company. What you will see are some examples of both internal and external creations. Some models work and others don't, because they didn't need to. As the Spin Pop® matured, the need to have fully functional samples decreased. Most of the first prototypes with characters had static figures only. Shortly thereafter, many prototype handles developed motorized articulation. Future samples would reflect this by having moving appendages. Most prototype samples were not connected to the motor for motion. The articulation was a manual feature and was good enough for the buyers to know what to expect in the real article. Most of them were used to the idea that the pop spins and the limbs move in the final production articles. This acceptance gave Cap Toys the ability to move quickly on concept models. The sales force would be armed with all sorts of new "what if" handles to sell, while at the same time marketing efforts would go after and secure the new licenses for production.

Original inventor prototype, the genesis of the entire line of handles.

Santa Claus & Snow Man. Note that Santa was resculpted prior to production. Shown here is the original pattern in same color scheme. Notice the corncob pipe in the mouth of the Snow Man figure.

Buzz Lightyear©. Prototype model is almost exact to production piece in every way. Notice blue wings and light blue for the sky painted behind head.

Syvester© and Tweety©. Decoration sample shown with moving arm and rotating lid. Notice the prototype label attached to handle.

Twin Spin Pop® prototypes. Left is the Hong Kong tooling model. Right is a hand built sample by book Author to prove that the concept worked.

Drill Pop® trio. Top is inventor model. Center is Hong Kong tooling model. Bottom is production shown with yellow label. Notice the transition from the first version on into production.

Life Savers Spin Pop®. Original fully working prototype model shown. Notice that the roll of candy was switched from horizontal.

The Little Mermaid®. Acetate pattern used in making of the tooling. This shows the resculpted face for running change in production. Shown without tail section.

The Little Mermaid®. Original model for decoration approvals. No moving features. Notice the facial detail. This was the first released face, which changed while in production.

Disney© trio. Canadian released assortment. Decoration samples shown for approvals. Notice the rearward facing action button on Tigger©.

Small Soldiers® pair. Decoration samples shown with posed arms. Notice the downward pointing dagger was changed in production

Godzilla©. Decoration sample shown with posed arms and jaw.

Wally Warhead©. Internal sculpted pattern for first time license review.

Batman©, Warner Bros©. Decoration sample shown on newest handle design. Cape is made of soft plastic material and not fabric as in the past.

Cat Dog©. Internal sculpted pattern for first time license review. Hand crafted by book Author for internal review meeting on property.

Sulley© and Mike©, Monsters, Inc®, Disney/Pixar© Decoration sample shown on newest handle design. No character motion features.

Harry Potter©, Warner Bros©. Pre-production samples shown without decoration. It is molded in various colors from plastic left over in factory for testing of tool.

## What If

Attempts to get popular and different properties were made by the research and development, sales, and marketing departments, and by the president of the company himself. Specially sculpted models had been made on speculation and hopes of securing the rights for certain properties. A special interest figure would be made and applied to the handles' mechanism. Most of these samples were one-of-a-kind and never reproduced. Usually, after the licenser meeting and the decline of rights, no additional samples would be made. What you see here are truly the most rare Spin Pops® ever made by the company. As seen here, they had never been released into production. Some of the properties eventually did make it into production.

One example of a sought-after license that eventually did go into production is for the Star Wars® films. In attempts to gain rights for the classic characters, a prototype of Darth Vader® with a chopping saber was made and presented to Lucasfilm Ltd.© This prototype is most probably lost. It was not until years later, for the release of Episode One©, that a Spin Pop® was produced with such a license.

Other attempted versions have been made that do not have a license and were never released to production. An example of this is a partner tool for the Drill Pop®. The idea of building a boys line called for another item—a power screwdriver with forward and reverse spin. It was made to expand the appeal of this boy-specific line. A working prototype was presented to the company, but never made it into production. The location of this prototype is also unknown. This would have been the very first Spin Pop® to feature a reversible-spin motion. We may see it someday in the future as a "knock-off" from a competitor.

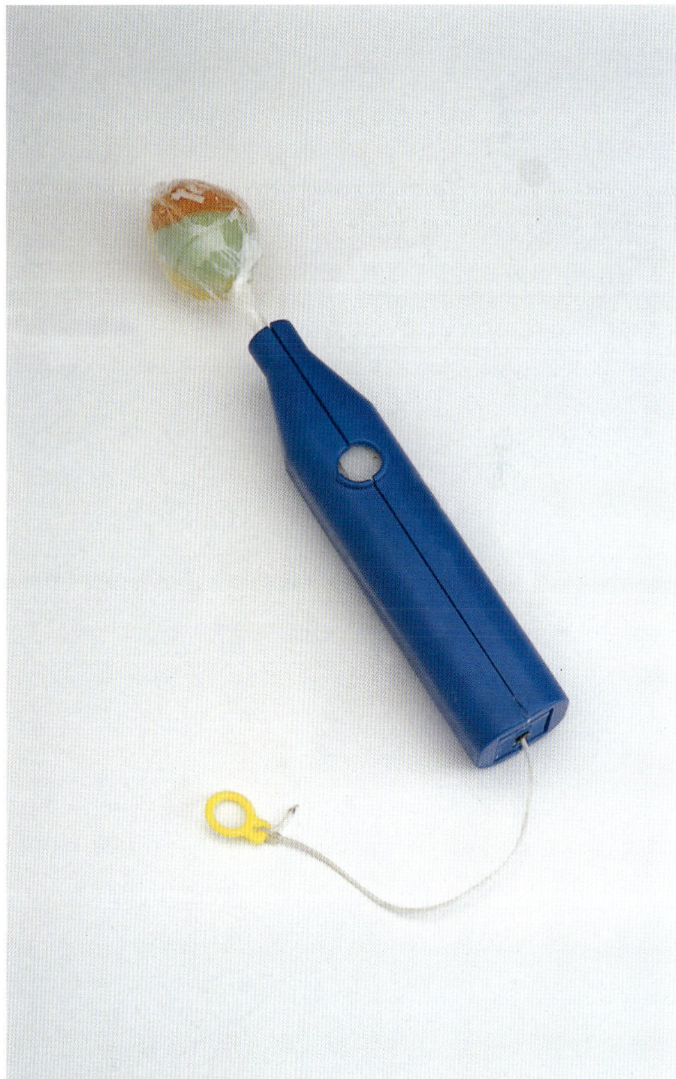

Hand powered pull string version of original handle. Pop spun when the string was pulled out.

Spin Jordan. Michael Jordan, Chicago Bulls©. Dimensional character of the basketball professional with spinning pop exiting out of right arm. Shown in team uniform. It is believed this model was crafted in order to meet the star player personally as well as attain the licensing rights. No moving features.

Barbie©, Mattel. Sculpted attempt for licensing rights to the popular girls icon. No moving features.

Mr. Freeze©. Sculpted attempt for licensing rights. Would have been an additional character to the Batman© line assortment.

Superman©. Sculpted attempt for licensing rights. Would have been an additional character in the DC Comic© line assortment.

Jar Jar Binks©, Star Wars Episode One®, Tooling pattern shown with no decoration. Never approved for production. Figure has wagging tongue and waving hands motion.

Wato©, Star Wars Episode One®, Tooling pattern shown with no decoration. Never approved for production. Figure has flapping wings and up and down body motion.

US006074266A

# United States Patent [19]

**Zak**

[11] **Patent Number: 6,074,266**

[45] **Date of Patent: Jun. 13, 2000**

[54] **SPIN POP WITH FACIAL EXPRESSION CAPABILITY**

[75] Inventor: **Marc Zak**, Sonoma, Calif.

[73] Assignee: **Oddzon, Inc.**, Pawtucket, R.I.

[21] Appl. No.: **09/243,894**

[22] Filed: **Feb. 3, 1999**

[51] **Int. Cl.**$^{7}$ .......................... **A63H 33/00**; A63H 33/26; A63H 13/00

[52] **U.S. Cl.** ............................. **446/73**; 446/236; 446/337

[58] **Field of Search** ............................... 426/104; 446/72, 446/73, 236, 337, 395, 484

[56] **References Cited**

U.S. PATENT DOCUMENTS

| | | | |
|---|---|---|---|
| 396,252 | 1/1889 | Beidler | 40/540 |
| 2,001,334 | 5/1935 | Stein | 446/384 |
| 2,733,547 | 2/1956 | Cohn | 446/391 |
| 4,279,419 | 7/1981 | Barnes et al. | 273/108.22 |
| 4,580,992 | 4/1986 | McKay | 446/321 |
| 5,209,692 | 5/1993 | Coleman et al. | 446/71 |
| 5,324,527 | 6/1994 | Coleman | 426/134 |
| 5,690,535 | 11/1997 | Coleman et al. | 446/236 |
| 5,957,746 | 9/1999 | Diamond | 446/236 X |
| 5,971,829 | 10/1999 | Hartman | 446/236 |

*Primary Examiner*—John A. Ricci
*Attorney, Agent, or Firm*—Marshall, O'Toole, Gerstein, Murray & Borun

[57] **ABSTRACT**

The present invention is directed to a candy holding device for holding a piece of candy and causing the piece of candy to be rotated. The candy holding device includes a housing, a drive system disposed within the housing, a rotatable member adapted to support a piece of candy so that rotation of the rotatable member by the drive system causes the piece of candy to rotate, a flexible member composed of an elastically deformable material, and an actuator coupled to the flexible member to cause a portion of the flexible member to move as the rotatable member rotates. The flexible member may be provided in the form of a face having a mouth area, and the actuator may be provided in the form of a connecting rod having a first end shaped in the form of a pair of lips and a second end operatively coupled to the rotatable member. The lips may be coupled to the mouth area so that the lips, the mouth area, and/or the cheeks of the face move in and out in synchronism with rotation of the candy.

**20 Claims, 2 Drawing Sheets**

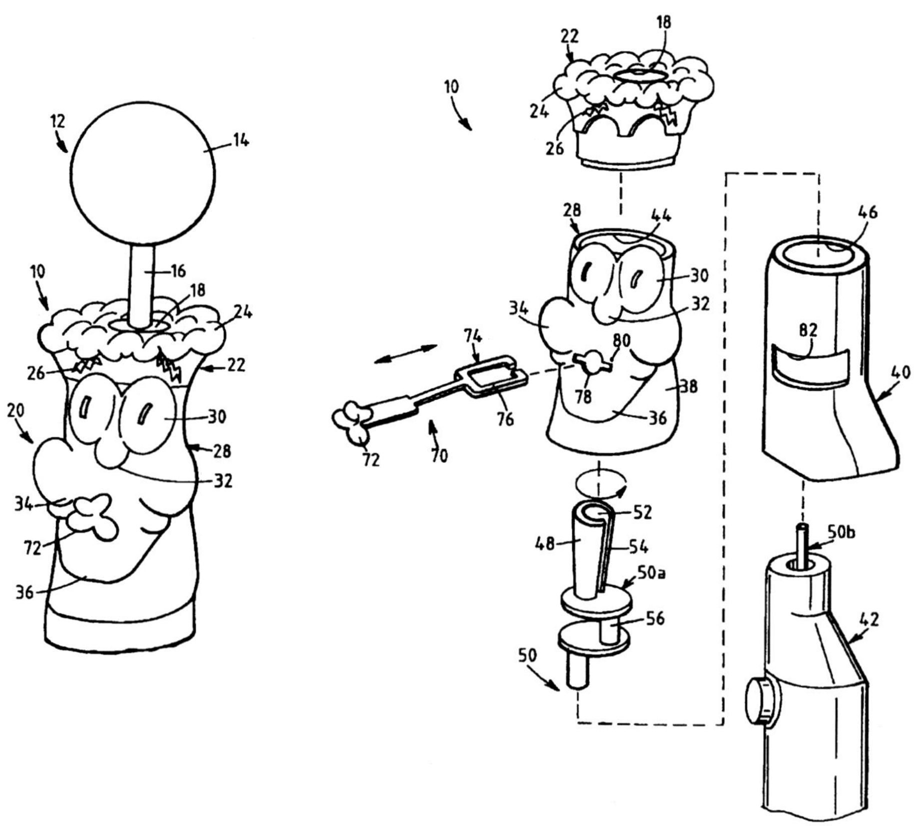


United States Patent # 6,074,266 Date of Patent: June 13, 2000. Spin Pop® with facial expression capability. Applied to the Wally Warhead© and Baby Dill© characters.

# Knock-Offs

The Spin Pop® is protected by several United States Patents. You can view the actual patents and the claims by logging onto the Internet. There are several search sites set up in conjunction with the U. S. Patent Office. You can begin a search with the words "Spin Pop®" and "patent." Some websites will allow you to search by the patent number. If you look on the handle label, you will find a patent number printed on it. The first patent number is 5,209,692. There have been several additional patents issued for enhancements to the handle over the years. An example of such a patented advancement is illustrated here.

The Spin Pop® was originally thought of as a novelty item, it was never expected to become such a huge success. Low sales volume in the first couple of years would support this thought. The belief at Cap Toys then was that international protection on the original item was not warranted. International business for Cap was limited mainly to Canada between 1993 and 1996, a length of time that let the application rights for international protection expire. This gap provided opportunity for others to make their own versions.

One of the first companies to copy the item was Chupa Chups operating from Barcelona, Spain. Their version looked very similar to the basic handle. Some changes to set it apart featured a dual spinning action of a pop and character. The little figure was attached to the side of the handle and rotated when the pop did. Others modeled their versions almost exactly after the original

Pop Machine, Chupa Chups. Spanish Manufacturer. Spice Girls© license.

Spin Candy, Chupa Chups. Spanish Manufacturer. Spinning figure off to the side of handle with slide on/ off switch.

Pop Machine, Chupa Chups. Spanish Manufacturer. Shown in blister card package. Notice the change in handle configuration. It is believed this design was copied by Hasbro for the newest Spin Pop® handle. In essence Hasbro is knocking-off themselves by copying those who copied them.

Chupa Chups hand powered spinning pop, Spanish Manufacturer. Just another way to get around the Spin Pop®.

Motor Mouth®, Mike and Jack, Australian manufacturer. This is a direct copy of the original handle made by Cap Toys.

handle. Many global markets never knew about the Cap Toys Spin Pop®. Australian manufacturer Mike & Jack, makers of Motor Mouth®, made the bold claim of being the originators of the spinning pop. Japan even joined in with both motorized and hand-cranked versions. They, like the others, would copy what had been done first in the US. Assuming that what they copied would do well for their market, Japan would even reproduce a version of the EZ Freezy Spin Pop Shop®. This was one of the more unique adaptations of the handle. It also was one of the lowest selling volume products ever released by CAP. French manufacturer Spel would become the largest offenders to knock-off the Spin Pop®. One look at the line they sell and you can see. Besides the physical handle design and function, they copied the trademarked Spin Pop® graphics with swirl pattern. The creation of unique variations, whether from Cap or someone else, was all done in attempts to expand interactive candy into a bigger business.

Action Pop, Rix Products, US manufacturer. Another hand powered spinning pop, but this time with a license.

W pop candy, Japanese manufacturer. Dual action spinning pops driven by yet another hand powered crank device. Notice green swirl label on a blue handle with orange colored knob on the crank arm.

Spinning ice pop set, Japanese Manufacturer. EZ Freezy Spin Pop Shop® want to be. Licensed version for international market.

Fairy Pop, Essential Box Co. Ltd., English manufacturer. Princess Pop® want to be. Up and down arm motion. Shown here missing wand in right hand.

Lolly Pop Drill, Essential Box Co. Ltd., English manufacturer. Drill Pop® want to be.

Space Gun Pop, Spel, and French manufacturer. Drill Pop® want to be.

Jump Jump Pop, Sunny Co., Chinese manufacturer. Shown in blister card package.

Jump Jump Pop assortment. Sunny Co., Chinese manufacturer. Licensed characters with up and down pop and arm motion. Note that pop does not rotate.

Jump Jump Pop assortment. Sunny Co., Chinese manufacturer. Smile face characters with rotating pop and arm motion.

Halloween Tricks pair, Spel, French manufacturer. Static figure with rotating pop.

Disney© licensed pair, Spel, French manufacturer. Static figure with rotating pop.

Spin Foot trio, Spel, French manufacturer. Soccer themed basic handle with rotating pop.

## Unique Variations

The interactive candy business was growing and with it the need to come up with new and clever ways to expand the Spin Pop®. Lollipop rotation would be applied to a variety of concepts. Both hand-operated and motorized examples had been prototyped and even produced. Handles would see innovation beyond the basic design and dimensional figures. Many of the adaptations departed from the general shape and design of the original Spin Pop®. What would link these cousins to the original item and protect the identity was branding them with the Spin Pop® name. Examples of this are found in the Wrist Pop®, EZ Freezy®, Life Savers®, and Twin Spin®. The Drill Pop® item does not have the Spin Pop® identifier, but is clearly a member of the family. Worth noting is the Twist Pop®. This item is not motorized, or even branded as part of the family. One look at it, though, and you can tell it is born of the same inspiration as that of the Spin Pop®. Of all the variations made, the one thing they all have in common is a rotating lollipop.

Wrist Pop©, a company creation. Wrist worn item that spins a small round sucker when the flip open dome is released. Shown in package with three extra Spangler suckers. Available in three different color schemes. Suggested retail $3.99. Cap Toys, Ohio. 1994. $10

Adventure Wrist Pop©, Inspector Gadget®, Disney©. Wrist worn item that spins a small round sucker when the flip open dome is released. Shown in package with three extra Charms mini suckers. Basic Wrist Pop© with new colors and packaging scheme with the addition of a compass to the dome top. Very limited release quantity due to only one retailer carrying the product. Very rare item. Suggested retail $3.99. OddzOn, California, 1999. $10-20

Twin Spin Pop®, a company creation. Two small suckers rotate at the same time when the button is pushed. Shown in an assortment of colors in and out of packaging with Spangler pop. Suggested retail $2.99. Cap Toys, Ohio. 1994. $8

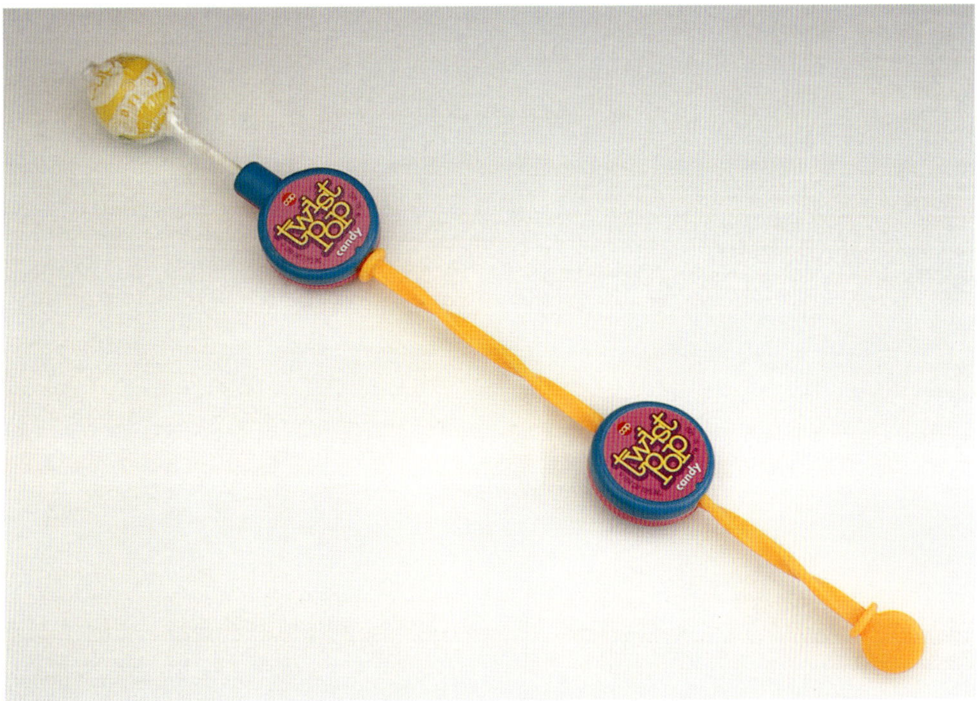


Twist Pop®, a company creation. Hand operated spinning pop. A close cousin to the motorized version released by the same company. Shown out of package with Spangler Triple Treat pop.

EZ Freezy Spin Pop Shop®, a company creation. An activity set for making your own frozen suckers. Mix up the sugar powder provided with water and then fill the molds. Insert the pop stick and place in freezer. Once frozen, place stick and sucker into one of the two spin handles and eat. Shown out of package with all accessories included. Notice the handle design with frozen buildup molded in. Suggested retail $7.99. Cap Toys, Ohio. 1995. $10-15

Drill Pop©, Boys specific product. Shown out of package with Unicorn Pop in two colors. First time the handle took the shape of a commonly found product. Second issue label in black. Suggested retail $2.99. Cap Toys, Ohio. 1996. $5-8

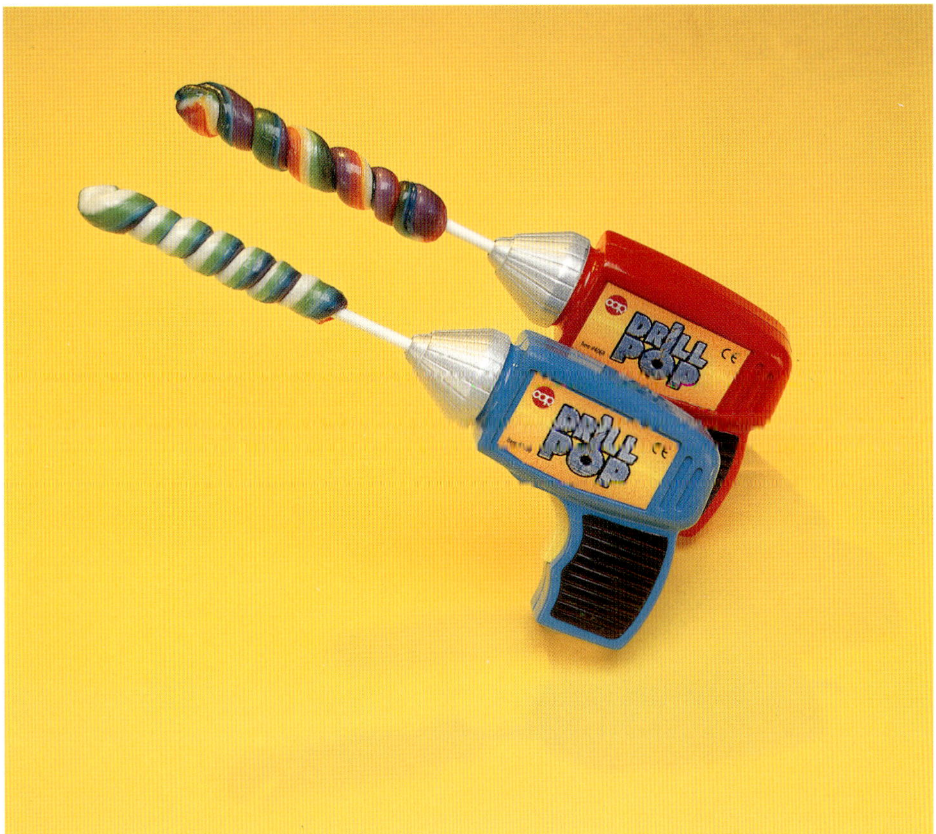


Drill Pop©, Boys specific product. Shown out of package with Unicorn Pop in two colors. First time the handle took the shape of a commonly found product. First issue label in yellow. Suggested retail $2.99. Cap Toys, Ohio. 1996. $5-8

Life Savers®, Nabisco©. A company creation with a branded candy. Shown in blister card package. Motorized rotation of an entire roll of candy. Vertical stacking of candy held by a center shaft, which is then connected to the handle. OddzOn, Inc. Ohio. 1998. $5-8

Spin Pop® Keychain, licensed product. Miniaturized keychain collectible that really works. Smaller than normal size with a pop that spins when the action button is pushed. Third color release of purple in blister card package. Suggested retail $4.99. Basic Fun, Inc. 2000. $5

## Fantasy Creations

One feature that sets any collection apart is the possession of a unique item you can talk about and show off. Hopefully, all the contents of this book will help you identify some points of interest for your collection. In addition to a rare prototype or an international handle, there are special "handmade" or "fantasy creation" handles to be found. The practice exists of talented craftsmen creating unique handles. The value of these rare and unique pieces is based on what you are willing to pay and your desire to have it. Most are generated for the purpose of resale. These are obviously not company-produced prototypes, thus the value is hard to calculate. They do represent the ability to enhance your collection with something most others won't have, and some of them are very interesting. Included here is an example of a fantasy Spin Pop®. If you encounter a prototype portrayed to be a company original, the seller should be able to establish its origin before you pay too much. If the seller cannot document a history, you may have exposed it as a fake. It's then up to you to decide the value. Most company prototypes are of high quality and made by professional sculptors and model makers.

G. Simmons and A. Frehley, Kiss©, home made creations. Altered from existing handles and decorated to match characters from the band. Listed on Internet auction. *Courtesy of Dennis Yount.*

## Exposed

As a collector you will want to know what to look for in determining a few things about a handle. Listed here are some identifiers of subtle differences between handles of the same property and mold. Many handles are being reproduced as they were first released. One way to date a handle is to look at the year on the label. Another way is to look at the inscriptions on the back of the figure.

For example, Buzz Lightyear© was reproduced from the 1996 molds for the release of the second movie in 2000. Everything is exactly the same except for slight color variances and the inscription on the wings. Some have the year inscribed on the figure, and almost all handles have a year printed on them. Unless a label is updated, the year on it will be when it was first released. Disney© handles are an exception to this; note that a Disney© handle's date usually relates to the year a property and character were released.

Taz© is a perfect example of a label that has not changed over the years. It has been in production since 1995, and the label has been the same. The two factories in China that made this product also put their marks on it. You can see physical differences on the characters' vortex. Despite color changes, there are clear markings you can see. At the base is a factory marking. One factory inscribed the letters PB, and the other KS. This is the first time the factory put its mark on a Spin Pop®.

Look at the face of the original Batman© released in 1994. One version has an upward mouth and the other has a downward mouth. Both were released at the same time. The frowning face is no longer produced. The upward smile is the same that was released later with the addition of the Robin© character.

Back of Taz© figure showing different inscriptions and color from first release and second release. Note the factory markings at base of vortex.

Front of Batman© figure showing different mouths. One is upward and the other is downward. Both released at same time, the upward smile was produced in larger quantities.

Back of Buzz Lightyear© figure showing different inscriptions and color from first release and second release.

Sometimes manufacturers will produce a product, release it, and then make changes for one reason or another. Bugs Bunny© has an interesting story in this regard. The original figure was approved and set into the early production phase with an extended left index finger. The toolmaker made a change for production reasons and moved the position of the extended finger to the center of the hand. Unwittingly, the toolmaker's change gave the figure a disrespectful attitude of social scorn. Ultimately, the decision to remove the finger entirely was made as a running change. The practice of making a running change is common among toy companies. Nevertheless, some handles with the center finger extended may still be in circulation. Such mistakes produce unique and limited versions to add to your collection.

Front of Pooh© figure showing different umbrella colors released and label changes. Same pattern used first in Canada and then released in the US.

Left side of Bugs Bunny© figure showing finger position altered by toolmaker in Orient and ultimately removed.

Tigger© figures showing color and label changes made from exclusive theme park market on right to mass retail on the left.

Front side of Mickey Mouse© figure from exclusive theme park market. This shows right arm variations in shoulder attachment. The version on the left was changed to a higher position.

Original handle design grouping showing color and material exploration. Notice the version on the far right has Chinese writing and is glow in the dark. None were ever produced for retail in these materials.

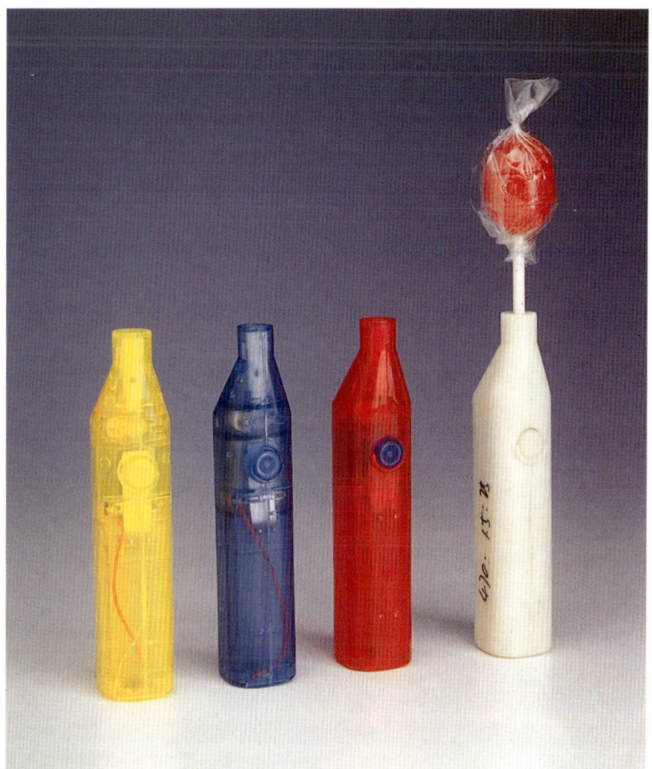

Buzz Lightyear©. Vacuum metal experiments in color for the second movie release.

# More Schiffer Titles

www.schifferbooks.com

**PEZ® Collectibles.** Revised & Expanded 4th Edition. Richard Geary. PEZ made it OK to play with your food, and now those imaginative, colorful candy dispensers are sought after by collectors worldwide.After PEZ candy and dispensers were brought to the United States in 1952, more than 250 different types of character heads were made. Licensed cartoon characters, movie and TV personalities, and original designs are divided chronologically into five series, usually indicated by patent dates on the dispensers. In this newly revised and expanded edition, most of the variations of PEZ dispensers are shown in 281 color photographs, as well as the premiums and store displays that helped sell them. The captions identify each item and give the current estimated values.
Size: 8 1/2" x 11" Price Guide 112pp.
281 color photos
ISBN: 0-7643-1095-X soft cover $19.95

**More Pez®.** Revised and Expanded 3rd Edition. Richard Geary. The popularity of PEZ candy dispensers and collectibles has never been higher. With the sharp rise in people suffering from "PEZ-O-Mania," RIchard Geary's classic reference is now bigger and better with new discoveries and new PEZ products. Illustrated with hundreds of color photographs (including many new to this edition), this book will contribute more knowledge to the field and serve as interesting source for building a collection. A newly updated price guide is included.
Size: 8 1/2" x 11" Price Guide 80pp.
170 color photos
ISBN: 0-7643-0994-3 soft cover $14.95

**M&M's® Around the World: An Unauthorized Collector's Guide.** Ken Clee & Joyce Losonsky. A hot, new collectible has gone international as great M&M's® items are being released all over the world! Now M&M's fans can see lists and values for all known M&M's dispensers, toppers, tins, and other items, plus related ads and displays from the USA and around the world. NASCAR M&M's items, French pocket surprises, and examples of earlier collectibles are all included in this expansive and colorful guide. Get in on the ground floor of this fast-growing hobby and join the hunt for "the collectible that smiles back at you!"
Size: 8 1/2" x 11" Price Guide/Index 160pp.
650 color photos
ISBN: 0-7643-1078-X soft cover $29.95

**The Collector's World of M&M's™: An Unauthorized Handbook and Price Guide.** Patsy Clevenger. Here's a sweet treat for collectors of colorful M&M's characters, starring regular and peanut. This is a serious collector's guide to those smiling candies with a tough exterior and a gooey heart. Author Patsy Clevenger, who describes the smiles of the M&M characters as "absolutely infectious," first provides readers with a concise review of the candy's history, including a timeline of M&M colors featured from 1940 through the 1990s. This comprehensive and enjoyable guide then leads the reader on a tasty tour of M&M collectibles, complete with values for the hundreds of items shown, which range from kitchen towels to tins, toppers, posters, and jewelry. Additional sections on advertising items and M&M packaging round out the book.
Size: 6" x 9" Price Guide 128pp.
298 color photos
ISBN: 0-7643-0406-2 soft cover $16.95

**McDonald's® Happy Meal® Toys from the Nineties.** Joyce & Terry Losonsky. McDonald's toys distributed in the United States from January 1990 through mid-1998 are shown and identified. 900 color photographs of Happy Meal promotions, including boxes, bags, premiums, and advertising materials, and a definitive numbering system identify each item, and check-off boxes help you organize and document your collection. Brief historical comments chronicle the world-renowned McDonald's Corporation.
Size: 8 1/2" x 11" Price Guide/Index 240pp.
900+ color photos
ISBN: 0-7643-0673-1 soft cover $24.95

**McDonald's® Happy Meal® Toys Around the World: 1975-1995.** Revised 2nd Edition. Terry and Joyce Losonsky. Welcome to the world of McDonald's Happy Meal® collectibles. These two books(USA and Worldwide)present a thorough list and complete Price Guide for the thousands of toys and collectibles distributed with Happy Meal® boxes and bags. Together, they are the most authoritative references. Thousands of all-color photographs show the toys, boxes, bags, advertising materials, buttons, pins, and variations given out both in the United States (one book) and in all the other countries around the world(another book).The authors have established a definitive numbering system to identify each and every item distributed with an alphabetical/numerical listing, and a cross-reference superseding numbering systems in other books. This set of books gives a time line of McDonald's Corporation history, check-off boxes to help you organize your collection, and the catalog of the items distributed with a Happy Meal® in the USA and worldwide. Read along and enjoy the fun!
Size: 8 1/2" x 11" Price Guide 270pp.
1850 color photos
ISBN: 0-7643-1093-3 soft cover $24.95

**McDonald's® Happy Meal Toys® Around the World: 1995-Present.** Joyce & Terry Losonsky. Fast food authors extraordinaire Joyce and Terry Losonsky continue their around-the-world McDonald's odyssey with this expansive guide covering Happy Meal® toys given out during the late 1990s in over 114 different countries. In addition to a complete listing of